MATHEMATICS CLINIC

Teaching methodology and useful seminar guide for educationists and Mathematics teachers

AMUSA ABDULATEEF

Author of **MATHEMATICS RHYMES**

Published by:

Addin Resources Ventures

Ibadan, Nigeria

+234 805671 0944   +234 80321 55018

latlib222@yahoo.com   addinrv@gmail.com

Visit amusa-abdulateef.com for other books

ISBN 9781977553850

(c) Amusa Abdulateef (2017)

© No part of this book is allowed to be reproduced, stored in a retrieval system or transmitted by any means without the written permission of the author. Violator would face the music accordingly for the infringements on the copyright of the author.

PREFACE

Mathematics is a natural science and the main language content of the creation. From the scriptures, there are several quotes showing that everything is created with measures in order to ensure synergy and social engineering. For instance, the amount of rainfall and its distribution to places in a year must be within a limit set. The intensity of sunlight must be at a point to have modest and bearable temperature. In actual fact, the research into the scriptures and deep reflections on the creation and how Almighty Creator does everything with measures even the lifespan of all creatures with mathematics, this prompts me into researching on how to solve all crises bedeviling the world with Mathematics. The work titled **"Mathematics, a natural art and master-key to unlock and solve all crises**'. For the sake of the inevitability of the subject, Mathematics, in all the courses of studies without exemption, and my flair for the subject right from my primary school days, I delved into further researching and writing on the subject for the progress of humanity in all specialization and spices of life since no one can do without the practicing of the subject every day, literate and illiterate, science-based or art-based professional.

I could recall what prompted me more into writing on how to teach the subject in the four walls of classrooms and impact the children. These are references made in the book. In the late 80's when I was in secondary school, for instance, our Mathematics teacher, Mr. Oluwole Fagbola, boasted that only the students of 'Additional Mathematics' would pass the mock examination that he would use WAEC standard to mark. I was not one of the students who offered the subject but determined to prove him wrong. I prepared for the examination and at the end I scored 59.5 over 60 of the theory. He could not fault my calculations simply because of the followings- I never jumped the steps and methods, the works were neatly done and I proved each answer at the tail end. This was a feat none of the ex-students at final year ever done in the history of the school. Later, we sat for the final ordinary level examinations, out of over 160 internal and external students that sat for the examinations, only six had credit and above in over five subjects that included English Language and Mathematics. I was one of the lucky few. If we calculate the percentage of those who passed Mathematics from the released figure, it was less than one percent. In addition, since when I passed out from secondary school that I knew I had come of age, I never heard of a year when a developing nation, in this case, Nigeria, has 100% credit pass in Mathematics. The recent ordinary level result released showed that over 49% failed Mathematics. The problem does not start from the secondary school or can we attribute the failure en masse to the teachers and other stakeholders in the secondary schools not to talk of the tertiary students that are allergic to the subject, the problem started from the elementary school whose foundation was weak and nothing to write home about. Pupils develop hatred and allergic to Mathematics right from the primary school. This needs urgent correction and the need for this research-based book for all stakeholders in the school business.

This is the presentation of the seminar aimed at deliver at a school whose pupils were under-performing despite all the inputs towards bringing the best out of the pupils in internal and particularly external examinations. The aim of all schools is to have pupils that do not just excel in exams but are mathematics-friendly in order to withstand any chosen courses in the future.

Without doubt, imparting the Mathematics knowledge and skill from the elementary schools has been the bane for the churning out graduates on science-technology based courses from tertiary institutions. It does not take a skyrocket to teach right and expect anticipated brilliant result from the impartation of the knowledge through provisions of right inputs at the right places. And without mincemeat of words, all pupils must be taught to embrace the subjects wholeheartedly as inevitable to all courses of study at tertiary institutions. Pupils drop interest in the subject that is inevitable to all courses even the art-based courses.

To this researcher-author, it is puzzled to see the reasons the pupils are not finding Mathematics as friendly despite its simplicity to teach with all graphic illustrations that are everywhere and at beck and call.

In view of the above, it is a right job to write on how to teach Mathematics that we tagged '**Mathematics Clinic**'. This is based on the methodology used at open seminars at selected schools correcting all stakeholders in education environment and the contents shall be useful guides to open the eyes of the school managements, the Mathematics teachers, the parents, the pupils, the school managers, the government and the community where the schools are located.

DEDICATION

The book is dedicated to the pupils and stakeholders on Mathematics and other forms of education.

ACKNOWLEDGEMENT

My eternal appreciation always goes to Almighty God for the inspiration to work on the subject towards adding towards creating greater interest in the science-technology based professionals with increasing Mathematics-friendly pupils with the research-based and thoughtful guide for all the stakeholders in education particularly mathematics education. Other appreciation goes to the doorsteps of my spouse, a treasure of inestimable values to me for all her supports, moral, financial and spiritual, in the person of Kudirah Joy Oladipupo; my great friends like Olalekan Joel Awujoola, Principal System Analyst of Nigerian Defence Academy Kaduna; Biodun Jimoh Tiamiyu of Nigeria International School, Cotonou, Benin Republic; Iziaq Ademola Abdurauf (Zico entertainment); Oseni Jimoh Folayemi, formerly Chairman NULGE of Odigbo Local Government Ore, Ondo state.

I cannot just forget the inputs of the likes of Mrs. Tomilayo Laniya of Bloom Heights Foundation group of schools; Mrs. Felicia Modupe Adeleke of Nickdel group of Schools; Mrs. Anike Abe and her hubby, Pastor Abe; Alhaji Iziaq Adekunle Sanni and his spouse, Alhaja Rafat Idowu Kunle-Sanni.

TABLE OF CONTENTS

PREFACE

DEDICATION

ACKNOWLEDGEMENT

INTRODUCTION

CHAPTER ONE

1.1 WHAT IS MATHEMATICS?

1.2 WHO TEACHES MATHEMATICS BY TRAITS AND QUALIFICATIONS?

1.3 WHAT IS THE GAIN OF LEARNING AND TEACHING MATHEMATICS?

CHAPTER TWO

2.1 WHAT IS/ARE THE CHALLENGES FACING TEACHING MATHEMATICS?

2.2 WHAT ABOUT THE STAKEHOLDERS INPUT TO ENHANCE MATHEMATICS-FRIENDLY PUPILS?

CHAPTER THREE

3.1. METHODOLOGY OF TEACHING AND IMPARTING MATHEMATICAL KNOWLEDGE

3.2 INSTRUCTIONAL MATERIALS NEEDED TO TEACH MATHEMATICS

3.3 GENERAL METHODS OF TEACHING

CHAPTER FOUR

4.1 SEMINAR SESSION

4.2 INPUTS OF EACH OF THE STAKEHOLDERS

CHAPTER FIVE

5.1 CHALLENGES BEFORE STAKEHOLDERS

SUGGESTIONS AND FURTHER RECOMMENDATIONS

ABOUT THE AUTHOR

ABOUT THE BOOK

INTRODUCTION

"I woke up at around five after meridian. I washed my face and hands with a half litre of water. I drank one glass cup of water of about 100 centilitre. Later, I bathed with a bucket of water of about twenty litres. I cooked two cups of rice with at least two litres of water, add some pinch of salt and ate with the stew prepared with fish broth. After dressing up smartly for office, I strolled a few distance. Along the way, I procured some stationeries at different costs from the road side kiosks after few bargaining with the traders..." What comes to the mind of the readers? We all breathe, speak Mathematics and act the subject voluntarily and involuntarily. Every trader is calculating and estimating to make the target profits. Every strategist in offices work with statistical data for use. The 'land economists' who are surveyors cannot do without bearing and distance. They use different tools especially the total stations to have the centre of gravity for all measurements. On the bases of the shapes and sizes of the land, the soil texture and other features on the land, the surveyors could guide the engineers on the roads to construct, hydro-power generating dams, the house to build, the weight of engines in auto factories among other places where the profession expertise skills are inevitable. What about the quantity surveyors that prepare the bill of evaluations in different estimates on the basis of the materials that must be used for the designs of the architects? If the builders or the civil engineer decides to use lesser quantities and the specifications of quality in the estimated bills of the quantity surveyor, the structure would collapse before the anticipated time for rehab. A standing structure must last several decades before there could be the need for repairs. The same is applicable to roads that are constructed based on professional specifications as regards the quality and quantity used. If the recommended, by the land surveyor on the site, amount in measurement of thickness of asphalt being laid on a road is reduced as a result of selfish interest or graft in order to make illicit gains, the road would soon start to have potholes and the laid asphalt would peel off. One can imagine the loss of lives and property such could cause the nation and the international community. What about the clothes we wear or the shoes in different sizes and shapes? The makers or designers had done a great work of calculations in measurements. In the solution boundary crises, the use of land surveyor in collaboration with the historian is inevitable. Think of it, is there any profession that does not need the Mathematics knowledge and skill? I believe none, even the historians, the philosophers, the psychologists, the public administrators and the likes in the social-science based, courses use time, figures and certain measurements to have quality presentation. How would a leader make fiscal budget without Mathematics? A business owner could determine profit and loss by simple Arithmetics. Economic crises are solvable by administrators through the use of statistics and other Mathematical applications. Asking me how? The historians must tell stories with references to TIME and DISTANCES of locations where such history was made or the event had taken place. The administrators can never be successful without the use of facts and figures collated into data through a branch of Mathematics called statistics. Through these, the tools to collect incomes and ease works are doable by applications especially electronic apps. Ask the team of builders on structures from surveyor (land economics), to architect that draws the masterpiece to the quantity surveyor that produce the estimates of the materials and the civil engineers that build the structure with the artisans. One can see that the knowledge of Mathematics is a must for all professions. Every lettered and the unlettered person cannot do without all the primary applications of Arithmetic (Get details in the book from the stable of the author titled "**Mathematics a natural art and master key to unlock and solve all crises**")

On the issue under discussion, I was a visitor to a private primary school running group of schools to meet a senior teacher in Ibadan sometimes ago. In the course of discussion reviewing the Mathematics results of the renowned school, he bared his mind that the subject has been the headache of the school management for some years. I itched to know the positions and the steps the authority had taken to have better performance. He told me of the two recent past seminars the school had conducted with the use of experts on the subjects invited from a university. 'What has been the result of the seminars? I asked. He tried to speak his mind with sadness 'below average despite huge resources invested on the seminars particularly the quality of the experts that conducted the seminars at different times'. 'Okay, I think I can render a better help to the school' I responded. He looked straight at my face wondering how I intended to go about solving a challenge that was almost perennial and common to all schools. He asked me how I would go about it. He was made to understand that the proposed seminar would be in two sessions namely the theoretical and the practical approaches. The first part shall cover just 30 percent of the time and the practical approach takes 60 percent when the last 10 percent covers questions-answer session as an interaction session between the participants and the facilitator. In short, the two sessions shall open all the stakeholders that would be carefully selected shall find the seminar one to remember which shall turn the pupils into above average scoring at least 70 percent in Mathematics examinations- internal and external. And the school shall have bigger number of interests in the science-technology based courses in tertiary institutions. This is a gain to the nation at large. He was so happy that he took me to the coordinator of the group of schools.

CHAPTER ONE

1.1 WHAT IS MATHEMATICS?

Mathematics could be said to be the key part of the language of creation 'Be'. The Almighty used and is using the language of creation 'Be' with the Mathematical flavor thus 'Be or exist in this measure'. Everything, every creature and every event are dependent of measures. <u>No one can do without Mathematics</u>. Let there be intensity of sun in this measure. Let there be rainfall distribution to this and that places in this measure. Let this quantity and quality of semen fertilize this number of eggs to have this number of foetus for this number of babies that would have this number of quantity of blood, number, shapes and sizes of veins, head, eyes, ear, organs, senses, limbs, bones of this shape and size including weight and with these and these measures of features; of this number, shapes, quality and size and then deliver from be the womb of this woman at this period or hour of the day or night in this particular place and year. The baby shall live with these measurements of things in life for this predestined number of years with this number of alluring ornaments of this world at his or her disposal; shall eat this quantity and quality of foods, shall drink this measures of water and drinks from second one to the last second of lifespan; shall attain this goal in life at these periods and several predestine about the once unknown or nothing that has become something (substance). Great is the Creator indeed. Mathematician is the Almighty indeed!

Looking critically on how man uses or applies the subject like the Almighty, Mathematics could be said to be primarily started with different ways to add together and subtract. It is a subject those who like and hate does every day. All literates and the illiterates cannot do each day without doing one form of Mathematics or the other. All professions can be successful with the application of the subject Mathematics. Naturally, all creations are based on calculations. And the addition of multiples or lists of numbers would be best done by multiplication. The opposite of multiplication, just as subtraction is for addition, is division. In all fields of life, we apply Mathematics. Leaders that fail to apply the subject would never have good administration. A leader at the helms could be productive and offer quality leadership with the application of Mathematics. Let us look at a few instances events from home to define the word "Mathematics", a family of ten has ten cups of rice to prepare. They are of different ages, would they share the prepared rice equally on the dining? Of course, no! They (each member) must be apportioned with the size of each stomach to avoid injustice, wastes and for health reason. Doing this, there must be either addition to some or subtraction from some, division and or multiplication especially in the sharing of inadequate soup or meat or fish. What about a pupil who has a limited resources to buy two products (school needs) of the same (equal) value with an amount that can only pay for one? What should he apply? His natural instinct of course to identify the most pressing need between the two objects. Economists would say he has done a scale of preference but can he do that without aware the VALUE of the needs at first? To fix a PRICE for a product or service, there is need to CALCULATE and WORK out the prices to insert a profitable price. Every specialized professional in short need one branch of Mathematics or the other to be successful in their positions!

Aristotle was reported to have defined Mathematics as the "science of quantity". The logicists reduce Mathematics into whole or partial logic with inductive and deductive reasoning. Intuitionists see the subject as a natural subject that is developed from Philosophy of Mathematics just as every other subject has its own philosophy. Formalists define the course from the point of view of the symbols and the rules operating them like the gravitational rule and others formulated by Mathematics theorists. This could be a reason why Haskell Curry defined it as the science of formal system. In the philosophy of Mathematics, there are the assumptions, foundation and implications of Mathematics which provide an account of the nature and methodology of Mathematics and to understand the place of the subject in people's lives. Statisticians feel high that Mathematics is a subset of Statistics and most valuable to all disciplines. This is partly true and partly false. We can rightly argue that Statistics is a subset of Mathematics and never otherwise!

Candidly speaking, no data can be collected or collated for proper interpretation or representation without all the major tools in Mathematics that is, counting, measuring and calculation aside the use of graphs, charts, index, parameters, degrees or percentages and many other Mathematical parts or topics. (Culled from one of the works of the author **'Mathematics a natural art and master-key to unlock and solve all crises'**)

Mathematics is therefore summarized to be a subject of counting, measuring (estimation) and calculation of shapes and sizes. These activities are recorded to be statistical information for the users. The topics of the subject are interconnected as they are somehow related. For instance, multiplication and powers are related to indices.. Some are similar but not same in shape or look. An example are the quadrilaterals.

1.2 WHO TEACHES MATHEMATICS BY TRAITS AND QUALIFICATIONS?

The tutors are simply officially Mathematics teachers who must have knowledge about all the branches of Mathematics such as Algebra, Geometry, Calculus, General Arithmetic, Trigonometry, Statistics, Logic and Probability among others. But, all the unlettered persons in artisans, rural dwellers, maids etc also teach the subject by their actions. Give certain sum of money to your driver to fuel the cars, he buys right. Give the illiterate wife some money and she has right financial plan on what to procure at markets. Mathematics. At the elementary stage from the lower cadre class to the upper cadre class, the pupils are treated with rudiments of the subjects. They are by qualification for the primary school should not be less than a holder of National Certificate of Education major in Mathematics. Such could in addition be first degree holder and major in the subject.

Talking about the man for the job (traits), such who pursues interest in picking the course as a course of study should:

- a) Someone who has special interest and passionate about the subject. The passion for the job could spur interest in using 'invented' market show in action to teach the children. Pupils could be taught simple drama as if they are in market place haggling over a price. For instance, in the teaching of shapes and price, a buyer would ask 'how much is the cuboid shaped match per unit. The seller could answer 'each cuboid of match costs ten naira. I mean 100 kobo.' Buyer may add 'what discount would you give if I buy a whole pack?' Seller may answer 'there is 2% discount for units of pack and

5% for a purchaser of 5 whole packs. …. This kind of 'drama' at the outset of a class would keep the spirits of the pupils high and more concentrated to the lesson of the week.

b) A Mathematician must be able to understand all rules of each topics. It is not expected of a teacher to fail to answer concisely questions about the rules that are relevant in the teaching of certain topic. Assuming a pupil ask a teacher 'why is it that the angles on a straight line is 180 degree?'

c) He must be able to handle all Mathematical tools. Someone who is teaching pupil-driver must not say that he does not understand certain parts in the car. He must have knowledge of the manuals of all cars. A Mathematics teacher is not different, he must know what all tools are used for since they are used for one measure or the other.

d) Such must have passion to develop strategies towards imparting Mathematics skills without stress. When I was in school, I read and summarize a text of not less than 300 pages inside an exercise book of less than 40 pages. I used to identify and pick difficult questions in a reference Mathematics textbook to work on when I was the subject teacher on a part-time basis in my community. Such additional Mathematics textbook like Backhouse and Harwood Clarke were reference texts that I used in my secondary school days to have deeper interest in the subject.

e) Such must be passionate about the subject even at the instance of hostility. One develops more interests every day for the subject. I used to relate one topic to the author. I like teachers who used to solve questions that are set on the basis of relationship with other topic. For instance, the property of rectangle could be used to arrive at simultaneous equation and the area or the perimeter could be solved with any of the methods to solve such type of equation like completing the square, elimination, substitution or graphical methods of solutions. It is a passionate subject that could go that far.

f) Such must be able to sleep and live the subject in a way to simply imparting pupils and others. To build bigger interests of the pupils in the subject, the subject teacher should use illustration explaining the subject. For instance, 'As an assignment, ask the pupils the number of kilometres covered by the cars of their parents per week. Tell them to convert the distance in kilometer into centimetres. Tell them the usefulness of knowing distance and bearing as inevitability in such courses like surveyor-land and quantity.

g) Such who has interest in Mathematics must be able to innovate new things and ways to solve all problem challenges. During my school days, in the course of answering all the textbook questions after a topic, I used to discover simpler methods to solve problems with proves.

h) Such must be able to define the right learning environment to study and read or work out Mathematical problems and tests. Studies show that the subject would assimilate under quiet environment. It should be better be done at the morning periods when the brains are fresh and the decibels of noise in the community are still minimal.

i) A teacher of Mathematics must be versatile in the introduction and imparting the knowledge. He should also be creative in the making, handling and the use of instructional materials, Mathematical sets and events within the school environment and the community of location.

j) No pupil likes dull moment in the class as the class should be charged to avoid dull session. The teacher should be a type that is jovial, be full of life always, articulate and a type of professional teacher who could come down to the level of the pupils being taken the subject.

k) In the recent, professionals who are specialized must be registered and should be the one that must pick up the job at the school. Like in other climes, the subject teacher must also be tested periodically to know their suitability for a particular grade or class. A teacher that failed grade six syllabus should be made to handle lower grade that he or she is able to pass the test excellently. In football parlance, performance on the field determine who make the first eleven or the referees to oversee a match. Referees are promoted or demoted based on performance. Teachers too should be promoted or demoted based on the class performance of the pupils.

1.3 WHAT IS THE GAIN OF LEARNING AND TEACHING MATHEMATICS?

Learning Mathematics has endless importance and relevance to the existence of all worlds. It has been predestined that every creature must use certain time to exist with measurable spices of living. In short, the inevitability in all aspects of life makes the embrace of Mathematics as a subject that must not be hated by all and not just the pupils. Mathematics should be loved and admired by all. All formal and informal schools should be right places of learning the subject. Parents should encourage the children to admire the subject through discussions and references at all times. Schools that do not have adequate provisions for the teaching and learning the subject is not worthy to be certificated for operation. In fact, all teachers in an education environment must have the subject as the second course hence they all appreciate the subject and designate to take the subject as roaming teachers at the absence of the specialized teachers employed to take the subject. The teacher must not be employed just to fill the vacancy but must be recruited based on the high level of passionate for the subject. Every subject is respected by all for the significance to life. The gains of learning and teaching the subject are extremely huge. Some are:

a) Learning improve the skills in measurements and calculations
b) Improves the ability to run viable business
c) It is an art that is inevitable to solve all crises facing mankind and nations (Ref: '**Mathematics, a natural art and the master-key to unlock and solve all crises**' from the stable of the same author)
d) It enhances the balance of the trades with ease for the managers of national economy
e) All courses of life cannot without the flavours of Mathematics. Historians use dates to back up their releases.
f) Teaching help to understand the topic the more

g) Teachers of Mathematics are the producers of all science-technology based professionals in particular and inevitable in other courses of study.

h) Teachers of mathematics are the moral and prudent people. A leader that is worth his onions must be sound in learning the application of the subject. It would help in the allocation of resources to ensure effective and efficient of manpower and resources.

i) Policy makers and users cannot do without the applications of Mathematics.

j) Mathematics assist the business and non-profit institutions to have right allocation of resources and measure the viability

THE FACT SHEET

We are all teachers of Mathematics regardless of our status, age, associations, level and types of academic attainments, gender, religion affiliations and ethnic values among others.

CHAPTER TWO

2.1 WHAT IS/ARE THE CHALLENGES FACING TEACHING MATHEMATICS?

In conventional schools, there are glaring challenges serving as obstacles to the teaching the subject. These are numerous namely:

a) Inadequate relevant easy to understand and interpret text books

b) Inadequate instructional materials that are majored on the subject to take all the pupils between the time frame allowed by the tutorial timetable

c) Improper organization and designation of subject teacher. A situation where a teacher that cannot handle more than ten pupils is made to handle fifteen would end up in low scoring or below performance children in Mathematics.

d) Unavailability of training on and off the course. It is expected that the teachers must have their degrees of knowledge of Mathematics and teaching the subject upgraded with updates.

e) More dullard and Mathematic-phobia are negative challenges to the impartation of the skills on learners.

f) Failure to share notes by the teachers as a result of self-pride and esteem. Just like the security agencies, the Mathematicians must compare notes. Once a teacher understands how to impart on pupils certain topic in comparable to others, the teacher would not have cold feet from asking others since no man is an island.

g) Inability of the teachers to plan ahead of the class. They lack pre-class tutorials by self or from others. A well composed teacher who had done the necessary before he enters a class would have confidence to stand before the great kings and queens not to talk of the young pupils.

h) Failure to enhance motivation of the specialized teachers. A motivated teacher of Mathematics would surely provide his or her optimum best at all times. Such would not mind motivating the pupils under them with gifts in the course of imparting the knowledge.

i) Inability to flow at the level of the pupils under him.

j) Inability to maximize what is provided as instructional and textual materials to achieve the set objective. Some teachers are simply wasteful.

THE CHALLENGES OF TEACHING MATHEMATICS AT A GLANCE OUTSIDE FORMAL SCHOOL SETTING

These are:

a) the teaching environment
b) the learning tools

c) lack of teaching skills by the use of right language for communication
d) the teaching methods informally at all places
e) the supportive teaching materials
f) the lack of legislation to support the teaching subject
g) lack of sensitization of the public on the importance of teaching the subject

2.2 WHAT ABOUT THE STAKEHOLDERS INPUTS TO ENHANCE MATHEMATICS-FRIENDLY PUPILS?

The stakeholders are formally targeted at conventional schools at the three tiers of education institutions and the first question shall be 'who are the stakeholders? We are all the stakeholders with different inputs to ensure right outputs. Some of these are:

a) The school and its management authority of board of governor
b) The staff of the school particularly the teachers
c) The pupils in the school
d) The textbook publishers and suppliers
e) The librarians
f) The parents or guardians at homes
g) The communities and the revered community leaders where the schools are located
h) The philanthropists

Each of the above has distinct features for the distinct roles to place down as inputs. Followings are the inputs:

a) The management must separate the Mathematics classroom from others as it assimilates at environment where there is no distraction.

b) The staff especially the teaching staff must be ready to put their all. Admit whenever wrong formula, method, textbook, selected question is used to the pupils; all the right equipment that would make quality Mathematics classroom must be intact by the non-teaching staff managing the specialized class for the school.

c) The pupils must endeavor to be Mathematics-friendly. They should be able to have no fear to ask relevant questions in the class during lesson and must not fail to answer all questions under a topic. They could lay their hands on all other textbooks from other authors outside the recommended ones to have broader knowledge about the work. The pupils should be confidence enough to tackle questions on the topic they were taught in class. Self-examination for teacher to mark is a way to learn to know Mathematics. A serious pupil would follow the steps and methods up to the point where he could not go further. Teacher or fellow brilliant mate would take him or her up from then.

CHAPTER THREE

3.1. METHODOLOGY OF TEACHING AND IMPARTING MATHEMATICAL KNOWLEDGE

METHODOLOGY

Methods of teaching vary from teacher to teacher depend on the level of ingenuity and passion for the job especially the subject-Mathematics. Each has his or her own style of teaching and imparting the natural art, Mathematics. During my time, I used to prepare pre-class introduction of a topic ahead of the week when such must be taught. I used to have the first meeting for introduction of the topic and the reasons why the pupils must be carried along. If I intend to teach kinds of triangle. I could introduce them to all three sided shapes of different measures as produced by me as a pre-class rehearsal. Later in the class, we would measure and jointly discovered that the sides are equal for one, two sides are equal in another and al the three sides are not equal in length in the last. The use of different colours could be used to keep the memory lasting for the pupils. Then the names of each is listed as equilateral, isosceles and scalene triangles respectively. If that is what is learnt and known to all the first day, I had scored 100% success. The second day, all the properties of each that must be re-identified before the new class session shall be written down. Then, we start building upon the knowledge. Generally, to produce right methodology, follow these tips:

a) Prepare for the class ahead with right seating arrangement with the recognition of the level of interest, vision, brilliance and height of the pupils consideration

b) Avoid piles of books and materials that could send fear to the pine of the pupils. By psychology of pupils, many hate to see tons of textbooks and past questions on the teachers' table. Introduce to them reference materials that would build the knowledge gradually. Pupils accumulate the knowledge from simple to the higher topics and questions.

c) Always introduce topics with right motivating words, analogies and the motivating words with cheerfulness (full or high spirit) in order to lift the spirits of the pupils even among the most uninterested ones in the class. Make the class lively and interactive.

d) Let the pupils understand the inevitability of the subject, Mathematics. Let them be aware that everybody needs the knowledge of the subject in all aspects of life even the illiterate. There should be relevant illustrations that would support this stand. With this, pupils would create more interest to learn to understand the subject topic by topic.

e) Ensure the distinguished graphic illustrations passed through all hands of the pupils in order to have clearer pictures that would bring home the main topic being taught.

f) Break down the topic into sub-topics. Do not just write the topic on the board, start from right interactive conversation that would be lucid and concise.

g) Allow open-book tests for the pupils. Allow the students to have the answers and give them set time to solve the posers under a topic. Avoid overloading with tests and home lessons. Carefully select those questions that would cover all they need to cover as far as the modules or syllabus is concerned.

h) Teachers should use different symbols that have different colours or marks to make illustrations. The Mathematics teacher could effectively use colours to differentiate the plane shapes in order to help quick reminder during tests, periodical oral assessments and examinations.

i) In some cases, Mathematics teachers should use simple 'poem' acronym, anecdotes, playlet, day to day activities, recent events, deaf language among others to impart the knowledge especially in teaching formulas, theorems, areas, perimeters, statistics and properties of shapes- two or three dimensional.

j) The use of carrot approach could work in the teaching of some difficult topics. 'if you answer this correctly, I shall give you extra mark' or 'you will win a cash prize if you get 100% in this test' or 'the neatly done and the right steps and methods shall be awarded bonus marls'.

k) Studies show that Mathematics teachers who used to scold pupils or come to the class with stern face used to chase away the pupils or discourage being friendly to the subject and the teacher. Taking each topic of a subject particularly Mathematics, as if it is the simplest among the subjects is always a plus. Teachers must look friendly and never be harsh in and outside the class for easy approach by the pupils to ask for teacher-pupil guidance over the difficult areas encountered in the topics.

l) Always allow the self-trial of the pupils. 'This is do-it-yourself.' When they solve a problem and find it difficult to go further, such should be sent to the class for communal efforts for the teacher to know the level of understanding of the class.

m) If it possible, enter the class with charts and simple questions and methods to solve them outside what are in the recommended books. Studies show that many recommended textbooks do not have simpler methods and formulas to solve problems under certain topics. This goes to the extent that methods adopted in a textbook may be simple and straight forward compare to the recommended texts. I used to follow samples of the Science and Mathematics Teachers textbook on Mathematics to solve some problems I found it difficult to use in West African Mathematics textbooks. But not in all topics. Harwood Clarke Mathematics was my choice learning and teaching certain topics. In some cases, the mode of teaching samples under a topic in West Africa Mathematics for junior classes may be more informed than that of the senior classes. I could recall that I relied more on the middle class textbook to sit and make credit result in an external general certificate in Education at my intermediate class. It was sweet experience using low class textbook to prepare and sat for such external examination and passed in the midst of difficulties.

n) In addition, good Mathematics teachers are good planners for pupils and arranger on the basis of defects and brilliance. They are patient and vocal in the presentation. The use of unambiguous words and simple analogies and anecdotes are useful for teachers to impart the knowledge. I could recall how we were taught shapes. We were asked to bring matches box, sugar box, package of biscuit and cube of seasoning from home to the class the next day by our Mathematics teacher. The next day, we were made to see the differences in the products. Every property of each were taught with ease as all pupils were fully participatory. Another teacher taught us, through orange that he asked us to bring from home, the introduction to the topic 'the great circle'. And the measurement of the performance of teacher is the 100 percent success of the pupils.

Behind the scene pre-class works are done by the teachers before class. It is like testing all the items at the laboratory by the teacher before inviting the pupils. He must have full understanding of the pupils by their level of interests in the subject. There are approaches the teachers must take. They must adopt several dynamic styles. **Teachers**, especially the formal but not excluding the non-formal, taking the subject, **Mathematics**, should:

a) Learn before attending class for the day's lecture. Visit the libraries for useful materials that would simplify the topics for the weeks, in and out. In all the environments are relevant events that would be useful to teach the pupils. It is easy for a teacher to teach the pupils about cost and selling price through the daily market transactions which the pupils have witnessed at marketplaces and retail shops within the streets. All the daily practical marketing situations could be used.

b) Prepare and rehearse before you enter the class through the use of recommended texts in the final external examinations with the help of the referential texts. Simple illustrations should be used to educate the children. For instance, coefficient of a number is different from the powers of numbers. When two is in the power of three, it means 2x2x2 making eight while two as a coefficient of three would give a product of two and three making six.

c) Prepare ground for the class before class. Learn to solve complicated questions from the simplest one covering and connecting different themes together. I like to solve equations from variation point of view while in school. I used to solve simultaneous equation that are based on the property of rectangle and square. This is a way of testing the understanding of the polygons and their properties. Anyway, all topics in the subject, Mathematics, have relationship. It is onus on the teacher of the subject to do his homework on next topic preparing adequately for the class.

d) In the pre-class rehearsal, use the right instructional materials. We used to be called upon to bring along certain materials to school by our mathematics teacher prior the teaching of the topic the next week or day.

We brought oranges to school prior to the study of the great circle. We brought matches and sugar cube boxes to school before we attended classes for the properties of the quadrilaterals. It was not just the theory but the practical approach complemented the teaching and easy impartation of the knowledge.

e) Prepare several other relevant materials to teach the class. Share your thoughts and notes with professional colleagues at neighbouring schools who your students have friends and families to co-habit with. As you share notes, the pupils from different schools and classes would share theirs. A scenario where the teachers and pupils work in collaboration to prepare the materials for the next class shall be a good motivation to be present in the class. Any pupil that misses the class would have his spirits in the class and would not take much time to catch up after getting the opportunity again.

f) Develop anecdotes and simple analogies for use in the course of teaching the class. Classroom psychology must be studied including the timing of the lecture. Class is made up of different pupils with different level of intelligence and learning rate. Some are slow learners. Some are fast learners. A good mathematics teacher should be good psychologist that must understand the mood of the majority, if not all the pupils, of the class. My teacher of mathematics used to be funny in the ways of talking and use of arms and legs while teaching the subject. Sometimes, he used to come to the class with strokes of cane. At the latter period, we were scared and would not be participatory except by force. The option of coming to class with canes was a turn-off to the pupils. And we found the subject too unfriendly and difficult to understand.

g) Learn how to carry along the brilliant and the dullard in the class in the question posted to them and the ways to answer with humility and friendliness. During our days, we used to have tutorial of four or five pupils. In the forum, there would be a leader acting the teacher or instructor. He or she led the pack as our captain. In the tutorial, we used to review the past topics, work excessively and extensively on the tests and questions. Through the solving of the questions, in the recommended textbooks and the recent past questions from external examination bodies, we identified certain difficulties and these were referred to the Mathematics teacher for solutions. Senior colleagues who were brilliant in the subject could render supports too in some cases as a relief to the Mathematics teacher.

h) Always prepare carrot and never use the stick approach for the excelling pupils. Teachers should make the class very interesting all the times for the pupils to enjoy and be proud to be part and parcel. Class motivation is part of the approach to draw all pupils to trickle and present in the class and settle down before the commencement of the subject. The anticipated motivation is spearheaded by the teacher in collaboration with the school management.

i) Never start teaching from the complicated angles. Start simple and build the topic gradually. A good teacher must break down the topic into units or sections. He should start from the simple introduction. I used to give respect to my Biology teacher, Mrs. Tomilayo Laniya, of Bloom heights Foundation, who used to start the pictorial illustration of an organism from the round-shape nucleus, then she would continue to draw all the internal organ till the last external membrane covering the internal organs. It would be a thing difficult starting drawing from the complex stage to the simplest components. This is the position all Mathematics and other subject teachers must tread to have brilliant pupils that would excel in their studies both in the internal and external examinations.

j) Get assistant if and when there is a need from consultations from colleagues and throwing such to the students as a form of assignment. I could recall that in the tertiary institution, we proffered solutions to the problems from our consultant-lecturers. Pupils are sometimes exposed and conversant with better referential materials and textbooks from other schools that could be used to solve a problem hard to crack by the teacher. A good teacher learns as he or she teaches from the pupils, colleagues and the texts.

k) Always prepare Mathematics excursions, within and outside the premises or the town the school is located, to increase the interests of the pupils. All pupils like to learn through travelling from a place to another. Imagine what your pupils would learn at Mathematics departments of the tertiary institutions. Guide career talks from the head of mathematics department and the librarians within shall boost the friendliness of the pupils to the subject.

l) Engage the pupils in intra-class Mathematics contests every end of a topic in the syllabus to ensure that they understand the topic to a satisfactory level. There are questions and answers in products and several tutorials from abroad that could be used to test the assimilating and understanding ability of the pupils. School authority and the Mathematics teacher should shop for such materials to engage in the inter-and-intra class Mathematics contests. I have seen 'who wants to be millionaire' quiz whose questions are Mathematics-based. Different tutorials on different topics are on sale and could be sourced for online at affordable prices. There are compact discs and e-books on the subjects that can serve as self-tutors for the pupils and the teachers. The teachers could watch and recommend the most relevant and adequately packaged to the pupils. Such should be discussed with the pupils at spare time hence he would know whether they had interest in the work from other shores and what they had learnt in the intellectual property.

m) Arrangement through the pairing of the brilliant and the below average pupils in tutorials and intra-class contests shall help them to be above average for the least and slow- learning pupils. Brilliant pupils would add values to the dullard ones especially those ones that are brilliant and accommodating to fellows and interested to lead them towards understanding. Such must have been taught that '<u>teaching is part and process of boosting one's knowledge and more understanding</u>' according to the noblest of the mankind sent to all creatures.

n) Expose the pupils to different ways questions are asked and answered by giving them all the questions under a topic to do as home work. I found it a nice way to learn more about Mathematics when I understand what a question demands from me. For instance, 'simplify' means break down to the simplest term or form. I like to have similar ways questions are interpreted in the subject. If it is written 'solve', it could mean find an answer or simplify to the simplest term. **'Factorize'** means 'simplify by reducing to the simplest form with the use of factors. A pupil that does not know what they mean by 'factor' or 'common factors' may not be able to factorize. Also such that does not know how to find the Least Common Multiples (LCM) may find many **algebraic** problems. A pupil that does not have knowledge of ascending and descending orders may not find some parts of **Statistics** difficult. The **probability** as a topic may be difficult for a pupil's failure to understand and distinguish between what makes a **whole set** or **sample space** from the **probable outcome** or **subset**. The pupils must have adequate knowledge about reciprocals (antonyms) in Mathematics and the synonyms from the early stage of learning the subject. All these little topics should be major instructional materials for the mathematics class and laboratory or library if the school could afford them for the use of the pupils and the teachers. (Ref: **'Mathematics Rhymes'** from the stable of the author)

o) Always develop simple ways to solve problems through personal researches into right referential materials. Pupils like easy ways to solve problems aside the moderate ones.

p) Pupils must be taught on how to independently use referential materials in and outside the Mathematics class

q) Pupils must be able to identify Mathematical set and other tools and should be taught how to handle and use each for different purpose. Understand what use compass, protractor and others for under different topics.

r) Mathematics session must be held during the morning periods when the brains of the pupils would assimilate things

s) Be patient and consistent in the work presentation. Never rush. Mathematics requires step by step presentation.

t) Each topic has different instructional materials that must be produced ahead of class by the class teacher.

u) There must be dynamism in the approach to all topics to have more interesting approach. Teacher must ensure the class is always lively in the dynamic teaching and imparting the knowledge of Mathematics.

v) Never admit where you got it wrong. Learn from your mistake. A teacher under my management of a school was taken a class 'perimeter of circle' but mistakenly used the formula of 'area of a circle' to solve the problem. There was grumbling from the class and I got closer. Seeing the mistake of the 'graduate' Mathematics teacher, I eyed him to see me outside after stopping the pupils from murmuring. At the exit, I corrected him. I told him to just rub off the word 'area' and replaced with 'perimeter'. He did that and the class became live again. Had he failed to correct the mistake and forced the formula on the throats of the pupils, the latter would be wronged for life!

SUMMARIES FOR TEACHING TO SCORE 100% SUCCESS RATE

WHAT THE TEACHER MUST DO

- Invent relevant teaching methods for each topic breaking down long methods to simple steps for proper understanding
- Always psyche up the pupils and the use of close samples like using market experience to teach such topic like measurements, prices, statistics
- Mathematics teacher should score 100% success rate by producing 100% pass. And also must ensure the pupils are Mathematics-friendly like him or her. Passionate teacher is up to task for the challenge.
- Rehearse behind the scene. Carefully study all the methods to solve a Mathematical problem.
- Use reference materials and relevant instructional materials on the right topic. The use of references is the employing other people's way of imparting the knowledge. Never be Mr. Know all. Be a pupil while preparing for the class.
- Try to identify the weak points and strengths of each pupil. Assimilating rate of pupils differ. Identify the rate. Suffice to start any topic with simple samples. In two options, the noblest started with the smallest. Pupil cannot understand ALGEBRA without knowing how equations are formed.

- Try to proffer solutions to each pupil by treating past questions. In the course of solving past questions, new ways of approaching questions are devised.
- Engage in personal tutorial and for self-development to boost the skills. Be book worm. Use recommended texts for personal upgrades.
- Give class and home lessons to pupils. Never jump the syllabus. Follow the guides to cover what is to be covered per week. Use the listed tools and develop analogies and every day activities. At the end, never fail to give both class and home works to the pupils.
- Find time to work together with the pupils. Closeness to pupils is to create tie with them. Let them be free to ask all sorts of questions. Answer them politely. Never shout or shut up a pupil itching to get the message clearer. Come down to the level of the dullest. Think for the set of pupils and not much for the brilliant among. If you are able to turn the dullards to average pupil, 100% success rate is sure!
- Create time to critically mark the lessons. Identify all mistakes and how the steps are followed. Never be so lazy to ignore to critically look into how they follow steps to do the home assignments. All lessons should earn marks and credited.
- Plan to work on the psychology of the pupils. Use motivating words that are positive. Never say 'you can never understand this' 'you're crazy', don't you have sense to know this simple thing'. Speak them positive words that would create eagerness to understand.
- Learn to know the learning rate of the pupils. From the time being used to do the class work, teacher would know the learning rate. Be patient enough to know each by the learning and absorbing rate.
- Arrange the pupils by the level of ear and eye strengths. Those with defects should be placed close to the chalkboard. Repetitions are necessary to help those who have ear defects.
- Learn how to teach the pupils on how to understand theorems, formulas and methods

- Learn how to teach the pupils on how to use Mathematical sets and instructional materials

- Study the degree of sight and the hearing deficiency of the pupils. Some pupils may not see the boards clearly while some may have ear defects. Consider their defects in listening and seeing.

- Engage the pupils on tutorial. Divide them into group. Let the brilliant lead the dullards to bring out average pupils in the latter. Create Mathematics group for the engagement of the pupils.

- Break the long methods to bits. It is onus on the teacher to devise ways to break into bit the long methods to solve a particular problem. The pupils must be able to know the steps in a concise manner for application to solve problems. Pupils should be taught how the formula is derived before the use at the first instance.

- Learn the class psychology. The general attitudes of the majority of the pupils during the class work should be known to apply the right teaching approach. There is always a need to generate full attention by carrot approach. Never use strokes of cane to force a knowledge on the pupil. No teacher is successful to impart knowledge and skill with corporal punishment via scolding, abusive words and harsh voice.

- Share intelligence on the topics. Share from your pupils as they are in the class to share from you. Share from your fellow Mathematics teachers within and outside your school. A knowledge starts where a knowledge stops. No man can say to be Mr. Know all except the Omniscient who is our Creator.

- Use the simplest methods. Simple methods are the first steps for the pupils to understand topics. Imagine telling pupils that the understanding division and multiplication gives way to indices and indicinal equation with proves on the chalkboard. They would learn to understand the foundation to be able to solve the bigger topics later

- Consult other teachers. Give no room to pride. As aforementioned, consult new and old textbooks. Ask others about how to go about a topic. Demand for the right tools that can be used to teach the pupils. These are all your

teachers that would help you scoring 100% success from the performance of your pupils at internal and external examinations on the subject
- Share intelligence on the topics. Asking others is sharing. Discussion at foras is sharing. Chatting with pupils after class is sharing with them. This is coming down to their levels to afford creating friendship and Mathematics-friendly pupils.

SUMMARIES OF WHAT THE SCHOOL MANAGEMENT MUST DO

- Mathematics class needs setting. The classroom must be set to speak the subject. The ROUND or SPHERICAL (OVAL) table, the walls must have Mathematics graphical illustrations having memorizable MATHEMATICAL RHYMES or PAINTINGS that have all the POLYGONS.
- There must be teacher that speaks the subject, devise right rhymes that can be chanted on measurements and relevant ones to introduce topics and
- The right and recommended texts and the reference books must be available at the beck and call. This is producing the right learning environment for the subject
- Provide right texts and reference materials
- Special class for sharing Mathematical topics
- Create Mathematics excursion
- Engage in Mathematics discussions on selected topics
- Open Mathematics inter-class and intra-class contests
- Expose the pupils into external Mathematics contests
- Train the teachers on duty
- Reward the pupils and the subject teachers
- Well stocked library with recommended and reference books
- Mathematics-friendly environment possibly a mathematics lab

- Train the trainers via on-and-off the job trainings
- Exposure to contests, inter and intra including the external ones
- Reward quality teaching and learning
- Non-performance of staff especially teachers must not be condoned
- Award to high performances in all classes
- Terminal seminar for the pupils and the teachers
- Produce right learning environment for the subject
- Provide right texts and reference materials
- Special class for sharing Mathematical topics
- Create Mathematics excursion
- Engage in Mathematics discussions on selected topics
- Open Mathematics inter-class and intra-class contests
- Expose the pupils into external Mathematics contests
- Train the teachers on duty
- Reward the pupils and the subject teachers

WHAT GOVERNMENT MUST DO

- Promote the subject as an inevitable and major to reading all courses via the supervisors of education
- Adopt right simple to understand texts and reference books as supplementary for schools and their stakeholders
- Institute awards and cash prizes for successful performance in all classes
- Ensure all schools have adequate teachers and assistant
- Special package in remunerations for the subject teachers
- Trained supervisors on the subjects to do periodical oversight functions to rate the teachers at schools

- Package relevant instructional materials for the private schools too as donation
- Support the private schools with grants or interest-free loans to improve aesthetics
- Re-write the curriculum or modules in a way for the topics to have synergy. A topic must be a preface to another. By this, pupils would not loss connections.
- Specialists in the subject should be those who would recommend texts and reference materials for the schools and the users.

WHAT PARENTS MUST DO

- Buying all relevant textbooks and instructional materials
- Invest on the child through hiring home-teachers
- Give 100% supports to the teachers, school and home through the procurement of the syllabus and monitor what is being covered time to time
- Motivate them through positive sentences like 'mathematics is the simplest'. 'The easiest subject is mathematics'. Use simple analogies to teach the pupils at home. Relate the home chores to Mathematics.
- Stock the library with useful and self-tutoring materials
- Select relevant texts and tools

N.B. We have used part of the brief hints with relevant experience in the course in the chapters so far. We shall further discuss the rest in the other chapters and sub-chapters for better understanding to the letter.

3.2 INSTRUCTIONAL MATERIALS NEEDED TO TEACH MATHEMATICS

In the course of proffering practicable solutions to the mass failure in the subject, Mathematics, all topics have peculiar instructional materials that would aid the anticipated performance. Teachers must have prepared ahead of class with self-made instructional materials that are not readily available in the school. Teachers could devise Mathematics Rhymes for some topics and formulas. The other needed materials are:

a) The furniture of different polygonal shapes
b) The objects of different shapes and sizes
c) The counters
d) The cardboards and pins to assemble relevant material in the class under the supervision of the teachers
e) Historical Mathematical maps and charts
f) Globe of the world
g) The cardinal points
h) Some types of measurements
i) Relevant textbooks
j) Mathematical sets for the class wooden or steel like
k) Breakdown of the Mathematical syllabus

Teachers and the pupils should create time to design instructional materials for the classes, offices (staff rooms) and homes. In the making of the instructional materials, they all learn some properties of certain shapes are known physically to the pupils. Through this, there would be easy memorization.

3.3 GENERAL METHODS OF TEACHING

Studies show that:

a) Pupils learn more from the use of illustrations and pictures with different colours
b) The hearing and seeing combine more to ensure understanding
c) Audible voice and the use of unambiguous words pass the message from the teacher to the pupil
d) The use of common events for illustrations is relevant in teaching.

Therefore, the teacher must:

a) Be loud and clear in the class. Pupils do not like a teacher that talks to himself in form of soliloquy. Teachers must always be aware of the pupils whenever in class. Wherever he stands, he should speak loud with lucid words.
b) Teach during the time when there is reflection of light. Do not teach at dark and near-dark places. Pupils must be able to see what is written clearly.
c) Teach in the class that should be situated at serene environment that is free from pollution especially noise or sound, mob of crowds and the related obstacles to classroom performance of the teacher.
d) Teach in class that is well ventilated with lights to ensure that there is no darkness which could prevent pupils with bad sight to see clearly. Teacher that has bad sights too need lighted class to be able to read, study and understand what is read from books and illustrative materials.
e) Always come to the classrooms with relevant events and illustrations that would help to deliver excellently in the class. Let the pupils be convinced and not confused about understanding a topic.
f) It is right for talking or illustrating with right items to be at least 75% while writing takes 25% of the periods. The 75% period of the time for teaching a

subject particularly Mathematics should include the questions and answer session as the teaching progresses, the pupil to pupil assessment as the class progresses.

g) Always give the pupils references as part of the assignment outside the school

CHAPTER FOUR

4.1 **SEMINAR SESSION**

The users of this work could take the seminar as a classroom exercise. This approach can be used for all other subjects. In this case, our focus is mainly on the subject Mathematics. And these are the steps to take:

a) The school must choose a right date and right venue to plan ahead. This should be done in collaboration with relevant expertise, educationists, publishers, professional associations and authorities and pre-seminar talks with the subject teachers. The most difficult topics in the subject should be the most focused upon. I could recall that many colleagues used to hate such subjects like trigonometry, geometry and some portion of algebra and statistics. Many students fail from the lagging behind in the memorization of the measurements by the pupils.

b) The right date should be a day that would ensure all selected participants from the school, the parents association, the subject teachers for all the classes, the selected pupils comprising the most brilliant, the average and the dullest in each of the classes, the community representatives, the education ministry representative, the board of management of the school and some observers must be attended.

c) The venue of the seminar must be of cross ventilation with adequate provision for the seat arrangement that must ensure convenience for all the numbered participants. The number of the participants must be planned for ahead of the meeting. The needed materials must be made available.

d) There must be adequate graphical illustrations for the seminar presenters for the use in the practical approach. If it is possible, there must be freedom for the participant to produce their own forms of instructional materials. The school could learn from the intelligent quotients of the participating parents, educationists, private instructors to add values to those of the subject teachers and the school. During my school days, we used to have different types of instructional materials produced from homes on a particular topic. And majority of the materials ended up to be useful for the teaching of the subject.

e) There must be a classroom setting for the practical approach for the pupils and teachers that are participating. The volunteers among the participants must be given chance to share their expertise knowledge on the manner to teach and impart the knowledge. I used to be teacher behind the scene in schools owned by people who are closed to me. The glory and praises are showered on the subject teacher when he or she is able to excel by producing quality products (high scoring and brilliant pupils) though assistance has been rendered at backdoor. It is onus for the teacher to throw aside the self-pride to embrace the better style to teach and impart the knowledge on the pupils.

f) It should not be working all day but a period for excursion within the school premises to show practical illustrations of what is taught in class. The teacher must have studied where to visit before the period is chosen. It could be done during the long term break of the chosen day. Sometimes, visit to mathematics library in the school during the break period under the supervision of the Mathematics teacher shall add values to the visits. School authority is needed to have the right stock of books and referential materials in the library.

g) In the seminar, the subject teacher and the pupils must be learners taking notes from all presenters. They should be free to ask questions as 'pupils' and never try to keep their questions. The answers to the questions from the teachers and the pupils could be the most helpful hints to ensure high scoring pupils in Mathematics in all examinations, internal and external.

THE THEORETICAL APPROACH

The need to look into the inevitable place of Mathematics in all day to day socio-economic activities deserve mentioning and the teaching by different situations cannot be over emphasized. The housewife that fails to apply Mathematics would be a spendthrift and may never cook right. If she visits markets for purchases, the traders cheat her through short measurements; if she fails to mind the pinch of salt and other seasoning in the soup, the soup becomes too salty; if she does not have the mind of calculation the heating period, she may burn the food on stove or cooking gas. She may administer wrong dose of pills to the sick baby for her failure to understand how to use simple arithmetic of measurement; she may administer pills at the wrong time. The road side vulcanizers despite being illiterate about Mathematics gauge tyres for motorists for a service charge every day. The unlettered traders in markets are working Mathematics every day to avoid loss of goods supply and money. Bus conductors and touts, that never attended any school for life, charge passengers by the calculation of the tons of loads and distance covered. Mathematics is not just the food for the Statisticians, the Engineers, the Pharmacists, the Surveyors, the Accountants and the Brokers. In short, all cannot do without mathematics regardless of their professions, status, qualifications and the rest. We feed on Mathematics. We rely on Mathematics for growth and development. We live on Mathematics. Can anyone dispute these facts?

THE PRACTICAL APPROACH

Set the seminar as in a classroom setting. As aforementioned, allow full participation of the attendees carefully selected for the seminar. There are better presenters among the audience than the facilitator invited to deliver the seminar by studies. All facing the writing board as if the students are facing the teachers in classroom environment. Starting from the brief introduction into the subject. Define Mathematics with some graphic illustrations used in the theoretical. Transform the illustrations into symbols, shapes and measurements by drawing them and use the measurements by symbols. Start the class by picking a topic which must be used to teach the subject as a demonstration indirectly teaching the subject teachers and the parents on how to take the subject at school and home respectively in order to create more interest to the listening and learning pupils. Our topic today is on MEASUREMENTS (Not really the topic in mind but should be written on the board boldly for all to see), (asking) 'What is the topic we are treating?' The audience representing pupils in class session would chorus 'MEA-SURE-MENTS'. As a teacher, I know you might be thinking of the area of measurements in my mind. MEASUREMENT is actually broad. We have measurement of distance, measurements of shapes and measurements of time among others. As a mindful teacher, I had to break the topic down to the level of what I intended to teach which has to do with the relationship between **MEASUREMENTS AND TRANSACTIONS**. Okay, write it down. Now, a family of a total weight 100kg must consume 15 cups of rice in a week, 20 cups of beans, 100 tubers of yams; each cup of rice at Bodija market costs … each cost of cup of bean at the same market is … and each tuber of yam at Sango market near the home and where it is bargained through window shopping to be the cheapest is …. Can we work out the total amounts needed for the food stuffs? The first step is to itemize the food items thus (starts writing them down one after the other by the demand multiply by the price per units. Then, what other cost would we incur? Chorus answer would be transportation and helper. Okay, let us assume … is the total cost of the two charges, then the total amount needed for the market transactions to bring home the foodstuff is … There should be consensus paper from the selected participants for the seminar to the benefit of the subject teachers and the pupils in attendance.

FACTS ON SYLLABUS

In order to have 100% result on performance of pupils in the subject, the style of teaching must be changed through the re-plan or the re-arrangement of syllabus. The syllabus must be re-written by the specialists in a way to have synergy of topics. Syllabus is to secondary schools what modules is to the elementary education. Both are always clear on issues:

    a) The topic and sub-topic to be taught the pupils in week

b) Such topic may be extended to cover more weeks for simple reason of allowing the content to be assimilated by the pupils as the content shall be applicable in major topics later in the term or terms

c) It is explicit in the type of approach of the subject teacher

d) The kind of instructional materials that are needed for the topic

e) The samples of questions to ask the pupils and the anticipated responses that would be part of the confirmation of understanding

f) The home lessons to be set for the pupils every day and for weekends

g) The class activities in the syllabus could be increased but the least in the contents must be achieved.

h) The teacher could use their intelligence, natural or developed skills from personal research or interpersonal relationship to re-draw or rewrite the syllabus. We do not mean that the teachers should alter the content but could simplify what to be covered in the syllabus. This is doable from the set objectives in the weekly notes. If, for instance, the syllabus sets limit to be covered to the use three symbols to do addition of numbers, the teacher should move up to add more symbols especially of those within the ambience of the school and at home to teach to understand.

Then after the class activities with all the issues clearly stated follow with touch of personal skills on how to enjoy better understanding, the teacher should give the pupils some class works and assignments.

4.2 **INPUTS OF EACH OF THE STAKEHOLDERS**

From the light of the previous sub-section, each of the stakeholders has special assignments. We shall do this one after the other.

SCHOOL AUTHORITY

a) Construct inviting and serene atmosphere in classroom purposely for teaching and working Mathematics. By a special class, let the furniture be in shapes and sizes including varieties of colour. The shapes of polygons speak volume about the algebraic aspect of the subject

b) Work in collaboration with the subject teacher to choose the right recommended books and referential materials for the pupils at different classes. The foundational classes must not be left out. In such classes of the crèche and playgroups, all the objects must be in shapes of sizes with simple Mathematical rhymes.

c) Through collaborative efforts of the subject teacher and the school authority, the syllabus or the modules could be upwardly reviewed for the areas to be covered to be unambiguous. This effort is aimed at developing teaching strategies that would work magic on the excellent performance of the target pupil-users of the contents of the syllabus or modules.

d) Just like laboratory rooms, there must be a separate Mathematics classroom or even called mathematics laboratory. There must a Mathematics library within the classrooms for all referencing and references by the teachers and the librarian

e) Create Mathematics-friendly teachers through adequate motivation in different forms such as the supplies of recommended textbooks, referential books, instructional materials, teaching tools and attractive financial motivation.

f) There must right ratio of Mathematics teachers to the number of pupils in each class. A situation where a teacher taking the subject is too overburdened to walk round the class, see the works of the pupils, correct each pupil and mark their loads of assignments does not augur well for the creation of mathematics friendly teachers and pupils.

g) The walls should have Mathematics-teaching wall charts as instructional material that speak directly to the minds and hearts of the pupils even a first glance

h) The authority must ensure that the subject is done every morning and not in the hot period of the day. Such class where Mathematics is treated should be in a ventilated with good shade that would enhance good breeze all the times.

i) The foundation class must be given all the right materials to teach Arithmetic hence the young ones should embrace the interest and love of the subject starts from the toddlers. The use of shapes and sizes including the counting of different measurements off-hand shall endear the interests from the level of education

j) The school should work hand in hand with the teacher to make pre-class arrangement for each topic for each week. The class should talk about the topic by the time the pupils arrive for the class in another week.

k) The Mathematics teacher specially recruited for the subject must be adequately remunerated and equipped with all textbooks required by the teacher themselves. There must be liberty for the teacher to choose from the lists of recommended Mathematics books from the education ministry.

l) Authority should ensure that competitions are held under inter and intra class Mathematics contests every week at least or bi-weekly. All good performances should be rewarded and the failed ones should be 'punished' for not revising their works and prepare well for the contest. I used to do this when I was managing a private school with the aim of producing quality pupils. Questions that captured what were done or covered in the syllabus would be set and each selected representatives of the class would pick from the list. Each correct score earned full mark likewise the bonus. We appreciated the winning class by prizes and carrots (strokes of cane) for the losing side). And the result was excellent. Within a month, the dull pupils moved up to become average ones. And the brilliant grows continues!

m) School management should expose the pupils to Mathematics clinic within the school under the specialized Mathematics teacher that must be employed by the school.

n) Pupils must be exposed to internal and external Mathematics contests within and outside the shores.

o) Efforts should be made for the school to operate separate Mathematics library with the support and collaborative efforts of the parents and other philanthropic organizations.

p) In general, all the ingredients needed to have quality learning environment must be produced by the management. Such includes paying for the on-the-job and off-the-job training of the teaching staff particularly the Mathematics teachers.

q) Create Mathematics hub in the school. Expose the students into inter-class Mathematics contests.

r) Register the outstanding pupils in the external contests. School proprietors could have inter-state Mathematics contests to build more interests for their pupils at various schools.

s) Award, promotion and recognition should be given the best Mathematics students on weekly basis for each of the class based on set assessments under the oversight function of the satisfied Mathematics teachers.

t) Excellent performing Mathematics teachers on termly basis should be highly recognized to boost his or her contribution to churn excellent pupils. Such recognition could be sending them to tourist places with his or her families on the bills of the school. It could be recommending for the next grade level in remuneration cash gift, plaques, enter the name in the hall of fame and special honour in an award night among others.

u) The authority could pool resources to fund research on how to create Mathematics-based sport in the shoes of chess and SUDOKU.

v) The teachers and the pupils should be partly funded by the management to have excursion to departments of Mathematics in the tertiary institutions. In the course, the heads of department of Mathematics would be good counsellors for the pupils to see Mathematics as inevitable and

THE TEACHING STAFF

The focus of the school and the parents is on the teachers of the subject. The publishers also expect the subject teacher to be able to take on the pupils with the contents in the printed works in a way the pupils would score very high in the internal and external examinations. The publishers used to follow the dictates and guides on the topics to be covered by the regulatory institutions as contained in the modules and syllabus for the subject. With the huge responsibility placed on the shoulders of the teachers, a lot of activities are expected of him or her for taking so huge task as a mathematics teacher. We have done this under the teaching methodology teaching what the teachers of Mathematics must do before entry into the Mathematics session class. In addition, there must be:

a) Demand for the syllabus or modules which would guide the teachers

b) Study what is expected of the topic for the period

c) Work to achieve what is anticipated within the set time frame

d) Never opt out of a topic until the necessary tests are carried out

e) Never relent to have right degree of understanding of your pupils attitudes to participating in the class, working and submission of the home-works, quick response to the class works

f) Never be furious with the management when the instructional materials demanded for are not supplied or undersupplied.

g) Politely demand for instructional materials from the pupils from homes. Teach and guide the pupils in the classroom under the classroom activities on how to produce manually certain instructional materials. They may develop better interests in the course of building theirs.

h) Carry along the pupils on how to identify and produce the instructional materials

i) Encourage the pupils to engage in tutorials with proper organizing of the tutorial on the basis of mixing the brilliant ones with dullard others hence the former could carry along the latter

j) Learn the level of understanding of the pupils' psychology for different class

k) Always study the class emotion and psychology before the commencement of the lesson for each day. Avoid transfer of emotion and aggression.

l) Plan to have the subject at the early hour of the day for the impartation to be excellent

m) Collaborate with parents and the school for supports that could be involved to ensure quality delivery.

n) Never transfer personal emotion and aggression on the pupil. Teacher must always use kind and words. Motivate the pupils over the gains and the inevitable need to be Mathematics-friendly.

o) Never be too arrogant to learn new skills to teach Mathematics. Just as you learn from fellow colleagues, endeavor to add to your natural endowment and personal ability in the understanding and teaching the subject, learn to learn from the intelligence and brilliance of your students who may have gone beyond others and you! A teacher was disappointed in himself when he was overconfident during a class when he failed to get the right answer to a question under the great circle. He was working one of the questions I had done privately outside the class. At the end when he finished without knowing that he had jumped a step. He itched to know from the class whether he got the work right. There was a dead silence in the class. He sought my view and I pointed it out to him that he was wrong. Imagine how the teacher was perspiring before he left the class sluggishly! In short, what you know how to teach, others may have greater ability and different styles to achieve better results at a shorter time!

p) Share notes with comrades from different schools. Never grow wings to ask from colleagues what you do not know the best method applicable to teach a topic. Such could be met during the on-the-job or off-the-job training course sponsored by the school or from self-sponsored training course. I could recall that one of my ex-Mathematics teacher in secondary school used to rub minds with the senior Mathematics teacher before she came to our class. In the course of sharing ideas, the teacher got the manner to pass the message (Mathematical lessons) better.

q) Never underestimate consulting widely as part of the pre-class work. Students like to be in class of a teacher that has confidence and enthusiastic on what they are doing. I developed interest in my mathematics teachers for his confidence and enthusiasm during my secondary school days.

r) Design Mathematics-based cards for the pupils for playing like ordinary cards. It could look like SUDOKU and other similar brain storming chess or cards.

THE PUPILS

First and foremost, the pupils should develop interest and love for the subject for its inevitable in all facets of life. Parents and teachers should collaborate with school and different stakeholders to have great admiration for the subject. A pupil that is anticipated to be Mathematics-friendly should do these:

a) Always create interests in understanding the subject since it is inevitable

b) Always engage in pre-class preparations before a class. Plan ahead the class with high spirit awaiting the class

c) Always think and breathe the subject. Be Mathematics-friendly. Since the subject is already communicated to be inevitable for all, the pupils must natural like the subject.

d) They should voluntarily form Mathematics club with list of activities that would improve their performance and interests

e) They must be Mathematics-friendly by being passionate about the subject. Talk the subject, walk the subject and do pre-class activities towards enjoying the class

f) Never indulge in the use of calculators and the formula table. Learn to know the property and how the formulas are derived. Let me give an instance. Perimeter has to with the addition of the shape. The area has to do with multiplication of lengths and the breadths. The two formulas look similar as multiplication is a short format to lengthy process of addition of figures. We can say that the summation of figures gives the multiplication of selected groups. Ensure you memorize the formulas and make sure you know the ways to simplify the questions. Learn how to interpret questions before you continue to work on it. By so doing, all steps would not be jumped.

g) All pupils must have tutorial class after the day. Self-tutor may not be enough for understanding topics. Involve in tutorial to share from the wealth of Mathematical knowledge and skills to solve problems from others. Do this with selected pupils hence there is no concentration of the brilliant ones. In the view of the above, no pupil should be allowed to use calculator. Use your brains and natural intelligence. I found out later about the relationship between addition and subtraction of powers of words in Indices and logarithm showing how to add numbers in powers together.

h) Ensure you memorize certain simple rules and logics. Memorization has been a major lesson we used to recite every day during my early education. We used to recite multiplication of numbers every day. Later the measurements off-hand. All these helped us to understand the subject from the primitive stage. The application does not mean to be a threat later. In fact, ask us anything in the measurements and multiplications, you find them chorused.

i) Devise the needed instructional materials for personal use. Just as the ones presented in the class, pupils could find producing self-made instructional materials in a replica of those at school at home. This effort shall help him to understand the lesson the more.

j) Be free to ask polite questions from the Mathematics teacher and your helping tutorial mate. Never shy away to ask about what you do not know. Never be arrogant to your fellow colleagues who have better understanding and Mathematical imparting skills even than your revered teachers and siblings at home

k) Learn to use simple formulas to solve topical issues on Mathematics. I believe that there are always shorter methods to solve questions unless the teachers or examiners ask otherwise. For a pupil, he or she must start with solving questions with simple methods and graduates to the complex ones.

l) The subject requires deep reasoning and patience solving of problem, always be patience and reason deeply. Never allow the calculators and the internet-based machines, because you are privileged to have them at your beck and call, to do the reasoning for you. Task your brain. Think over what the questioners' demand of you via having the question broken down into lines. Then, deduce what is anticipated of you.

m) Understand the questions before you start working. Read between the lines to the level of having right interpretation. This is buttressing the above. Never start to work on any topic until you have elementary understanding of the topic. Guide yourself with the tutorials and lessons from your teachers and fellow brilliant pupils. Critical look at the samples or examples in the textbooks and the recommended referential materials shall be of utmost use for your understanding and learning of the subject.

n) Always use referential textbooks to have better understanding of the topics including the discovery of the simpler methods to do the job. I used to use Harwood Clarke mathematics as referential book towards understanding topics in our syllabus some twenty five years ago. And this developed my interests and understanding on the subject. At the end, my ordinary level results was one of the best in my zone with Mathematics inclusive.

o) Use topic of the subject as nickname for all-time understanding. When we are in school, we used to call one another nicknames such 'Pentagon', 'Mr. Algebra', a friend now the Principal System Analyst in a high profile security institution in Nigeria was nicknamed 'ogive curve'. Through this, pupils are reminded the day such topic was taught.

p) Create adequate time at the right time to study the topic already taught in the class. Pupils could create special time for themselves to go over the topics taught in the class to have broader knowledge. Meditation over topics could bring fresher memory and understanding. Sharing the gains from meditation with fellow pupils of like-mind shall add values to the level of meditation. One could share the feelings and understanding with pupils from other schools to share from the intelligence of their teachers and fellow pupils too.

q) Have previews about the new topic before class through personal research into the topic through the use of referential materials. During my days, we used to have access to syllabus to know the next topic in line for the actual week. Students used to procure syllabus to know what is to be covered per term. By this we would know what topics to be learnt ahead and prepare for such with our senior students at school or at home. Through this, some difficulties ahead of the class are becoming a simple thing during the class. They used to share their past experience and how the subject or topic was finally understood without much stress.

r) Never underestimate your colleagues by height, skin color, speaking English language speaking fluency. There are quiet students who are not active in class and never participate in the interactions but are simply brilliant in Mathematics. Study to discover them and establish friendship with them. I had met in school friends who were poorly dressed, unkempt and not fluent in speaking English language but was very good in calculation and all topics in Mathematics.

s) Consult books from lower classes to have better understanding of mathematical topics. I learnt deeply about several topics from the consultation of books of lower classes while at my alma mater. In many cases, writers or authors of lower class-focused books used to use clearer languages and simple samples to teach a topic compare to the authors of the senior class textbooks. I learnt and understood more big topics from lower class textbooks than the ones recommended for my class.

t) Make special friend or establish close relationship with the Mathematics teachers. Ask your teacher after class. Never feel too shy to ask questions. Do not be afraid of being called a vocal ones when it comes to asking about the steps where you are lost. Teachers are there to patiently listen to the pupils and solve their worries on a topic. The joy of all teachers is to have 100% understanding and success rate. Always receive the guides and tips or hints from the teacher at the right time.

u) Practice what you learn from the lips of your teacher and tutorial mates and always follow the guides. After series of efforts in developing greater interests in solving different problems under all topics chapter by chapter in selected textbooks, I decided to solve all past questions covering at least ten years back. I had made attempt to solve all questions in Harwood Clarke until another greater project took its position.

v) Visit other classes being handled by another mathematic teachers to have a piece of the methods being used to answer certain questions. This could be a way of sharing notes to learn a great deal of wisdom on new tactics. Studies show that other teachers may have better and clearer explanations to solve a Mathematical problem.

w) Develop interest in the visit of places where Mathematics is directly and indirectly taught. Such includes seminars, workshops, symposia and the departments of Mathematics.

x) Volunteer yourself to lead group of students of your level to seniors who can add values to what you were taught by your teachers. This is a right step in the right direction. In my days, we used to visit our seniors for better illustrations and the pupils in lower classes used to come to me too. I found out that through teaching of others, one learns more.

y) Learn to make the best use of senses. Do not sit at the distance if you have sight problem. Never fail to ask question if you have hearing defect. It is right to ask if you do not hear and sit closer to the whiteboard if you are seeing clearly from distance. A pupil that keeps mute when he or she does not understand or see what is written clearly is contributing to his or her failure in examinations and lose of interest in the subject.

THE PUBLISHERS AND SUPPLIERS

A major stakeholder in the education industry is the Publisher who supplies the schools the recommended textbooks and other referential printed materials for the subject. A better research work on each of the topic in the syllabus must be collected by the publishers through their research and innovation department. Good Mathematic books, based on their working relations in collaboration of the team of specialists that prepared the syllabus and modules adopted with the regulatory agencies, for the pupils must be those:

a) With adequate graphic illustrations with different designs and shapes for each of the topics

b) Whose topics are to be inter-connected or have right synergy in unambiguous manner
c) Easy to interpret words- that is lucid and concise
d) Contain pictorial illustrations with relevant guiding colours at appropriate pages
e) No economizing of pages in the use of samples, graphical illustrations and tests. The clearer the pictures and illustrations the better the understanding for the eyes defective pupils. The more the samples from different angles and the tests, the better the imparting the knowledge by the teacher.
f) Anecdotes and analogies that are picked from the environment of the pupils being taught
g) Adequate instructional materials to teach each of the topics
h) Guides and hints of what the teacher and the pupils could do ahead of the class as pre-class activities
i) Topics arrangement in ascending order hence pupils can build the knowledge and develop more interests as they move from one topic to another.
j) Adequate questions set differently for the pupils and teachers for practice
k) Publishers could add values to the syllabuses and the modules by splitting the contents down with simpler sub-themes to be covered.

THE LIBRARIANS

a) Shop for the right materials for the library under your watch
b) Always be ready to assist the pupils ready to learn the subject. Do the work of counsellor to encourage them into be Mathematics-friendly
c) Work with collaboration of the school management and subject teacher to add value to the library and the users.

THE PARENTS OR GUARDIANS

They are authorities on their own. Without the parents, there are no pupils for the schools. Social responsibility on parents or guardians is huge towards producing Mathematics-friendly pupils. These would be done:

a) Procure the recommended textbooks chosen by the school management in collaboration with the Mathematics teachers after proper scrutiny at the walls of rooms at homes and offices for the pupils to learn from at all times.

b) Procure relevant instructional materials that would aid learning and understanding of the subject like the ones find in schools

c) Employ professional Mathematics teachers for your children or wards at homes or offices where they retire after school

d) Equip the home library with useful referential Mathematics textbooks and ensure the home-teacher use the material for the pupils.

e) Parents should support the school management and the teachers through rendering all supports, financial, moral and technical ones. By finance, the supplies of instructional materials and textbooks including referential books by parents and the association of parents in collaboration with state-government endorsed school governing boards

f) Parents should cooperate with the management to have special symposia, seminar, exhibition and workshop on the inevitability of the subject.

g) All the pupils must be monitored and adequately supervised the execution of the home assignments from the teachers to them. Necessary assistance should be given to ensure the pupils are Mathematics-friendly.

h) Parents should be volunteer counsellors especially on career where the pupils are made to understand the inevitable position of Mathematics in all courses of study at the tertiary institutions.

i) The parents or guardians should discourage their children from the use of calculators and laptops to solve Mathematical problems. It could be right for the undergraduates doing research-based works. At elementary stage of education, the brains are still fresh for deep thinking to find and proffer solutions to problems and puzzles. Instead of procurement of calculators and the likes for the pupils, the instructional materials that they could also interpret with their senses should be bought for them.

THE COMMUNITY AND THE COMMUNITY LEADERS

a) The community that host the schools should see them as part and parcel of development to the area as their children are the primary target among beneficiaries

b) The leaders in such community should carry along the others in the community over how the school could be assisted in the creation of serene and noise-pollution free environment. Studies show that Mathematics is a subject that would not be understood under tense condition. I never found all Mathematics class held after long break around 12.00 p.m. easy to understand despite my special interest and flair for the subject, Mathematics

c) The communal efforts should create Mathematics-based library and tutorials after school

d) The community should join the school management on its efforts to promote pupils-friendly subject.

e) The Mathematics teacher should be applauded and showered with encomiums after each terminal performance. The performance of pupils is enough for the encouragement of the teachers.

THE PHILANTHROPISTS

a) They should help equip the special library with recommended and reference mathematical books

b) They should institute awards and prizes for winners of mathematical contests for pupils of different classes.

c) They should at all times be in support of what the school authority and management deemed fit to raise the standard of the performance of the pupils in mathematics towards churning out graduates from science-technology based courses in tertiary institutions.

CHAPTER FIVE

5.1 CHALLENGES BEFORE STAKEHOLDERS

These challenges identified are solvable by taking the opposite side facing each of stakeholders one by one.

SCHOOLS

a) Short time for the teaching of the subject. Despite the compulsory of the subject as a core course, schools used to allot little time to teach and impart the knowledge on pupils. Topmost priority should be the subject for all schools. And the timing should be at the right period of the day particularly in the morning.

b) Incompetent of teachers handling the subject. In some cases, school employ teachers that took the subject as sub-courses are employed in order to avoid right remuneration for specialists that are most qualified and competent. What is expected of a computer science or integrated science/Mathematics graduate taking Mathematics in the schools? These are not specialists like those whose course of study is mathematics (major).

c) Lack of training on and off the job for the specialized teachers. Most schools do not see it as a duty to upgrade the knowledge and skills of their teachers especially the Mathematics teachers. It is a must for the proprietors and proprietresses of schools to organize periodical seminars for the teachers to meet and share notes on how to get the maximum results from their pupils. External facilitators and seminar paper presenters should be part and parcel of the seminar to expose the teachers to the latest methods of teaching and imparting the knowledge of Mathematics. All teachers that find teaching a particular topic in the subject would have learnt how to go over it when they resume at school at the end of the seminar sponsored by the school management. School owners could cut cost by inviting facilitator (expertise Mathematicians from relevant institutions and associations) to school premises over a weekend to engage the teachers and expose them to the new ways to take the pupils.

d) Lack of adequate of instructional materials to teach the subject. Schools could shop for different instructional materials from publishers within and outside the school base in order to boost the learning and teaching ability of the pupils and teachers respectively. In fact, the teachers would find it as easy as a, b, c to have clearly illustrated instructional materials. There was a time I designed unique instructional materials on Mathematics for supplies to schools through the State Universal Basic Education Board (SUBEB). Unfortunately, the project never saw the light of the day. The contents in the specially designed and prepared Mathematics-focused instructional materials are self-tutored for the teachers and the pupils and each could easily replace the textbooks except the need to treat the questions in the textbooks.

e) Non-cooperation of the parents with the teachers on the performance of the pupils especially in ensuring that all home lessons are done without copying one another or whole spoon-feeding. The nature of jobs of most parents do not pave the way for supervising the works they do at schools especially on the core subject, Mathematics. What is expected of a parent that leave as early as eight o' clock and arrive at eight in the night? Such would have no time for engaging the children and wards on what they were taught in schools particularly the Mathematics as a subject.

f) Failure of the parents to provide similar Mathematics atmosphere at homes and offices. Studies show that most parents do not mind the outcome of the brilliance. They are after the success at all cost and by all means instead of working in collaboration with the other stakeholders especially the school managements and the teachers. Many parents do not see it as a responsibility to buy right instructional materials for their pupils. Many do not procure the recommended texts not to talk of referential textbooks for their children or wards.

g) Lack of juicy gifts and awards for good performance serving as carrots approach. The class teachers should be recognized just as the pupils in the award ceremony at the end of the terms. Special recognition for the mathematics teachers would be a morale booster for the staff holding the subject in the school. If all the classes are based on performance of the pupils in percentage of scores and the number of the pupils, the teachers that are overall first to three shall be a motivation for the school. Mathematically, if the pupils that scored above 70% in a class is over 99% in a class, such point could be used as yardstick (cut-off point) to measure the performance for recommendation for the awards for the best three.

h) Lack of organized competitions on Mathematics for all classes of pupils especially in the primary school. The class pupils should be exposed to inter and intra class competitions on the subject. All contests organized by the business institutions and government should be participated in. the Cowbell Mathematics national contest used to strike my mind. There is nothing stopping schools to engage in international Mathematics competition. One can imagine the gains from a competition between school in neighbouring state with another or a school from Nigeria to another one in Ghana. All these exposures of pupils and teachers would assist a great deal for the students and the teachers to learn more from the knowledge of others.

i) Inadequate funds to prosecute the good intention of the school management to have 100% success rate. Most public schools used to have badly motivated staff and non-dedicated teachers. This is also a problem facing the private school owners. The capital from financial institutions are not satisfactorily for business survival. The loans from banks used to have high interest rates and other stringent conditions.

j) Inadequate supports from the government for the private schools towards performing the social business. Many supervisory agents from the ministry make abnormal and illicit demands in graft from the school and by so doing, reducing the funds for other purposeful usage.

SUGGESTIONS AND FURTHER RECOMMENDATIONS

Schools must organize seminars following the tips released in the work. The pupils and the subject teachers deserve to be updated and upgraded. The efforts shall be able to produce the professionals. Others are:

a) There should be general acceptability of the schools towards adopting seminars and other references suggested in the book to produce Mathematics-friendly pupils and graduates

b) The attitudes of the subject teachers to ensure the pupils are subject friendly must be positive and full of optimism.

c) Mathematics should be done at symposium and workshops for the stakeholders

d) The major stakeholders should play major roles in the making the subject number one.

e) There could be special schools for the subject right from the foundation or elementary schools under the management of the specialized professionals.

f) Schools should have inter-school mathematics contests periodically just as they used to have joint mock examinations for the out-going pupils. Such contest must be on the subject, Mathematics only.

g) Schools should have forum where the Mathematics teachers rub minds together periodically to discuss the topics and solutions to hard ones. The sharing of intelligence shall be a way to improve the knowledge about the subject and how to impart the knowledge on the pupils.

h) Psychologists and counsellors on teaching should be engaged to re-orientate the teachers on how best to take on the students in this internet-dominating world. Pupils should be discouraged from the use of simple machines like calculators and laptops to solve problems. Those machines are meant for the adults at work to lower their level of thinking on different issues. The only assignment before school-age pupils is to cudgel their brains on how to solve issues. After all, computers and all other simple machines being used are products of brains of certain inventors. What could distract the young ones to apply their brains where and when it is necessary at their prime? Studies show that most pupils rely so much on the use of computers and calculators to solve problems instead of tasking their brains. They prefer to designate responsibility to the computers to have easy answers to questions.

REFERENCES

AMUSA ABDULATEEF, Mathematics picnic, a seminar for school pupil, Addin Resources Ventures Ibadan, 2009 Nigeria

Ditto, Mathematics teen-contest for pre-secondary school pupils, Addin Resources Ventures Ibadan, Nigeria 2004-2007

Ditto, How to run tuition-free school (unpublished) 2008 Addin Resources Ventures Ibadan, Nigeria

Ditto (2017) School business, challenges, practicable solutions and prospects Addin Resources Ventures, Ibadan, Nigeria

ABOUT THE AUTHOR

He is a prolific and creative writer, think tank, socio-economic researcher, essayist, child educationist, activist, public analyst who has authored several **bestselling** books such as JOBS WITH ZERO CAPITAL (1& 2); CREATING NEW JOBS FROM THE EXISTING JOBS, BEING MY OWN BOSS; WASTES TO WEALTH JOBS;

HOUSEWIVES ARE PROSPECTIVE ENTREPRENEURS among others on **ENTREPRENEURIAL GROWTH AND DEVELOPMENT.**

He has also authored several supplementary books on socio-economic issues facing the nations such as WINNING HUGE SALES AND INCREASING CLIENTS BASE; ECONOMIC RECESSION, THE CAUSES, THE SPIRAL EFFECTS AND THE PRACTICABLE SOLUTIONS; UNDERSTANDING BUSINESS ENVIRONMENT, RIGHT OR WRONG; MATHEMATICS A NATURAL ART AND THE MASTER-KEY TO UNLOCK AND SOLVE ALL CRISES; PIRACY, THE TRENDS, THE CAUSES, THE SPIRAL EFFECTS AND PRACTICABLE SOLUTIONS among others.

He is also a playwright and literary genres writer. He is a product of the citadel of technological innovation, The Polytechnic Ibadan where he graduated with good grade in Business Administration. He is from the Alapa family from the paternal side and happily married with children.

ABOUT THE BOOK

Learning from the manner of creation by the Creator with the application of Mathematics, the author sees the teaching of the inevitable subject as inevitable. Constant failure of pupils in the internal and external examinations is a major reason for studying about the reasons for the failure of pupils. Studies of the author show that pupils are not Mathematics-friendly as a result of different factors such as poor teaching methods by teachers, poor recruitments of wrong teachers for the job, poor motivation of the recruited specialized teachers, poor exposure to modern methods of teaching, lack of instructional materials, poor and inadequate equipment for teaching the subject and either partial or non-participation of the stakeholders towards turning the subject into pupils-friendly subject. The book is drawn from a seminar paper presented to a school whose major challenge is how their pupils could be able to pass with flying colours in the inevitable subject, Mathematics.

The author used his wealth of teaching experience as child educationists and his ingenuity in the know ledge of Mathematics to conduct the seminar. The contents are written in the book for the users of all stakeholders such as the school authorities, the government educational agency and departments, the community where schools are established, the pupils and the other users of the Mathematical skills and knowledge. This simplified version serves as a prized asset to all the listed stakeholders.

www.ingramcontent.com/pod-product-compliance
Lightning Source LLC
Chambersburg PA
CBHW031548210526
45464CB00003B/1205